2nd Grade Math Workbook
Counting Money
Math Worksheets Edition

SPEEDY
PUBLISHING

Speedy Publishing LLC
40 E. Main St. #1156
Newark, DE 19711
www.speedypublishing.com

COINS

Write the value of money on the next pages in cents.

 1 ¢

 5 ¢

 10 ¢

 25 ¢

 50 ¢

_____ ¢

_____ ¢

_____ ¢

_____ ¢

_____ ¢

_____ ¢

_____ ¢

_____ ¢

_____ ¢

_____ ¢

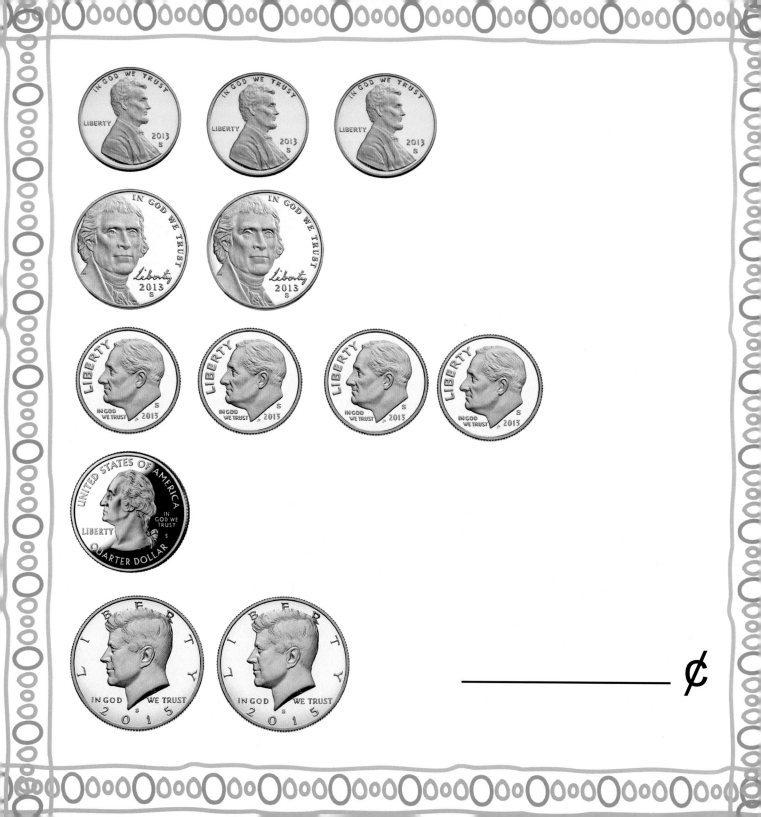

_____ ¢

BILLS

Write the value of money on the next pages in dollars.

$ _____

$ _____

$ _____

$ _____

$ _____

$ _____

$ _____

$ _____

$ _____

$ _____

$ _____

COINS + BILLS

Write the value of money on the next pages in dollars.

 1 ¢

 5 ¢

 10 ¢

 25 ¢

 50 ¢

$ _____

$ _____

$ _____

$ _____

$ _____

$ _____

$ _____

$ _____

$ _____

$ _____

$ _____

ANSWERS

COINS

121 ¢	186 ¢	124 ¢
194 ¢	129 ¢	176 ¢
196 ¢	148 ¢	178 ¢
145 ¢	94 ¢	

BILLS

$158	$11	$226
$87	$173	$48
$172	$47	$176
$232	$110	

COINS + BILLS

$75.55	$22.85	$26.18
$170.21	$7. 09	$11.25
$27.10	$111.11	$23.25
$11.21	$11.85	

96846160R00026